THIS WALKER BOOK BELONGS TO:

WHO IS THE MOST DANGEROUS CREATURE ON EARTH

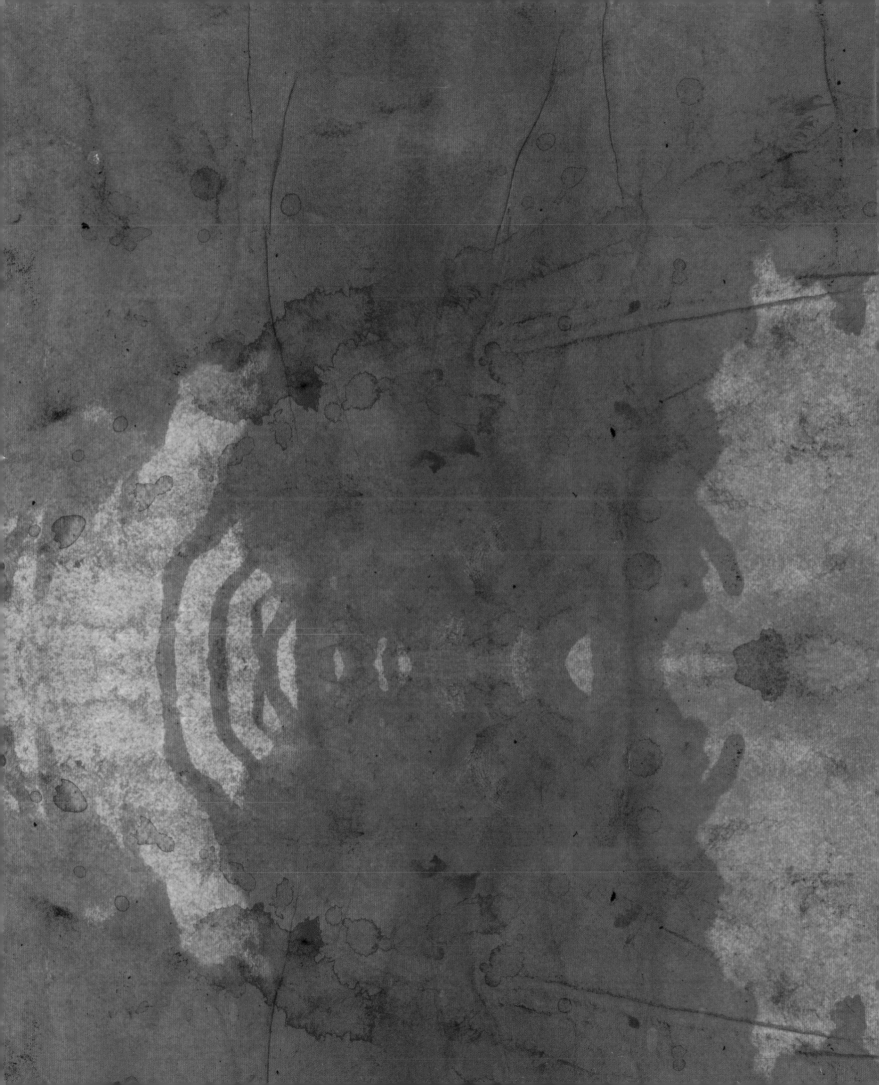

For Ben (who was so deadly with a magnifying glass) N. D.

With thanks to Lucy Ingrams N. L.

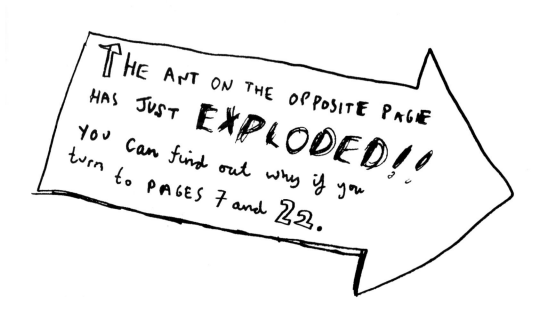

THE ANT ON THE OPPOSITE PAGE HAS JUST EXPLODED!! You can find out why if you turn to PAGES 7 and 22.

First published 2012 by Walker Books Ltd, 87 Vauxhall Walk, London SE11 5HJ

This edition published 2014

10 9 8 7 6 5 4 3 2 1

Text © 2012 Nicola Davies   Illustrations © 2012 Neal Layton

The right of Nicola Davies and Neal Layton to be identified as author and illustrator respectively of this work has been asserted by them in accordance with the Copyright, Designs and Patents Act 1988

This book has been typeset in AT Arta

Printed in China

British Library Cataloguing in Publication Data: a catalogue record for this book is available from the British Library

ISBN 978-1-4063-5742-4

www.walker.co.uk

Deadly!

The Truth about the Most Dangerous Creatures on Earth

by **Nicola Davies**

illustrated by **Neal Layton**

WALKER BOOKS
AND SUBSIDIARIES
LONDON • BOSTON • SYDNEY • AUCKLAND

ANIMAL ASSASSINS

Stabbing and strangling,

poisoning and drowning,

electrocuting, exploding, dive-bombing and even death by gluing!

No, this isn't a description of a horror movie, but some of the ways animals kill each other.

In fact, when you look around the animal world, it's clear to see that animals have been almost as good at finding different ways to hurt and murder each other as we humans have.

☠ ATTACK!

African lion
grabs prey with 5 cm claws then stabs with 8 cm canine teeth

HDR ☠ HIGH

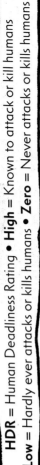

HDR = Human Deadliness Rating • **High** = Known to attack or kill humans **Low** = Hardly ever attacks or kills humans • **Zero** = Never attacks or kills humans

Mantis shrimp
hits prey as fast as a .22 bullet

OW!

HDR ☺ LOW

Peregrine falcon
dive-bombs prey at 200 kph and stabs with talons

 HDR ☺ ZERO

Poison-arrow frog
has skin containing deadly poison, which kills anything that tries to eat it

 HDR ☺ LOW

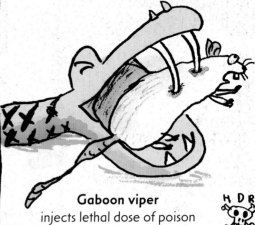

Gaboon viper
injects lethal dose of poison through 5 cm hollow fangs

 HDR ☠ HIGH

Carpenter ant
explodes to cover enemies in glue, killing them by suicide bombing

 HDR ☺ ZERO

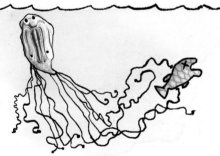

Box jellyfish
has 60 tentacles, each with thousands of venom-packed stinging cells

 HDR ☠ HIGH

Bombardier beetle
sprays boiling toxic liquid out of its bottom

 HDR ☺ ZERO

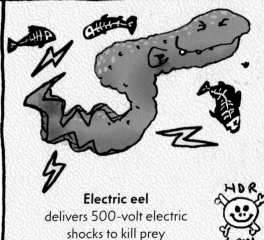

Electric eel
delivers 500-volt electric shocks to kill prey

HDR ☺ LOW

Spitting cobra
squirts venom into the eyes of animals that threaten it

 HDR ☠ HIGH

Of course this kind of violence in the natural world shouldn't surprise us. Many animals are predators – professional killers that must kill prey to survive. Evolution has equipped these creatures with deadly weapons such as teeth, claws and poisons, plus speed, stealth and strength.

But predators don't have it all their own way. As they have evolved better ways of killing, so prey animals have evolved better ways of escaping or fighting back. The result is a planet full of animals that are armed, dangerous and deadly!

KILLER CATS

Big cats are the top predators in habitats across the world and are kitted out for killing. They often work by night when their eyesight is six times better than our own, so locating unwary victims is easy. Big ears help gather every sound and sensitive noses detect tiny traces of prey. Padded feet and camouflaged coats get them close enough for attack. Claws – up to 5 cm long on every paw and kept razor-sharp by sheathing them when they aren't used – act as grappling hooks, to latch onto prey and pull it down. Then, big cats have a mouth full of butchery tools to deal with their unlucky dinner: two pairs of canine teeth, which are like daggers and up to 8 cm long, for delivering a killing bite;

premolars (called carnassial teeth) shaped like scissor blades to chop flesh into bite-sized pieces and a tongue like sandpaper to rasp meat from bones. Huge jaw muscles give big cats a vice-like bite and prevent the jaw from moving from side to side, so struggling prey can't wriggle free.

All cats – even little ones – have this death-dispensing equipment, but different species use it in slightly different ways. Most are lone hunters using stealth and camouflage

A SMALL CAT (WITH THE SAME DEATH-DISPENSING EQUIPMENT)

PRRR!

TOOTS

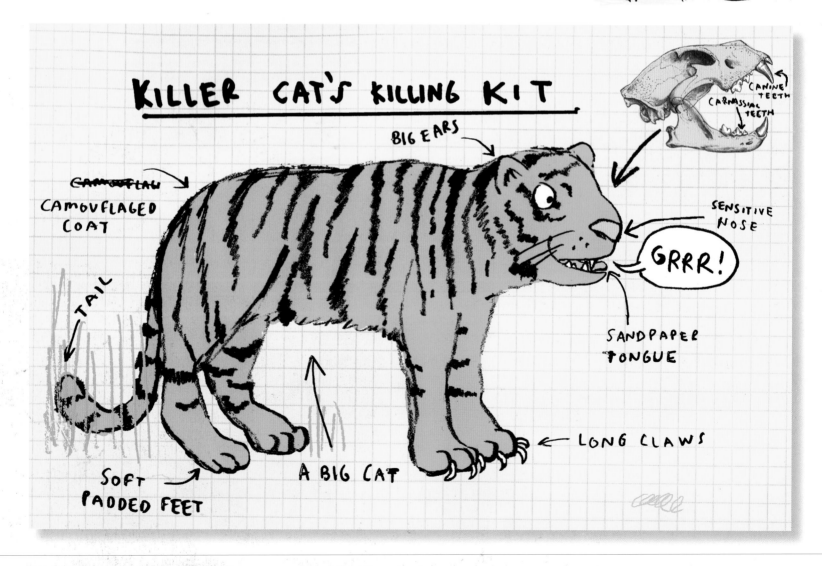

KILLER CAT'S KILLING KIT

BIG EARS

CANINE TEETH

CARNASSIAL TEETH

CAMOUFLAGED COAT

SENSITIVE NOSE

GRRR!

TAIL

SANDPAPER TONGUE

LONG CLAWS

A BIG CAT

SOFT PADDED FEET

A TIGER

to get them close to their prey, then a quick pounce and a stabbing bite is all that's needed to get their dinner.

The biggest cat of all, the tiger, has stripes to provide camouflage, and once it's grabbed its prey – a deer or wild pig – it delivers a bite to the throat or the back of the neck. Its smaller South American cousin, the jaguar, has a coat of spotty splodges to hide it in the jungle, and its super-strong jaws allow it to kill with a skull-crunching bite to the back of the head.

A JAGUAR

CHOMP!

A SUPER STRONG LEOPARD

Like jaguars, leopards have spots to help them blend with their background, but as they share their habitat across Africa and India with tigers and lions twice their size, they need some extra tricks to cope with the competition. First, they are super-stealthy and can get within 2 metres of prey without being detected. And second, they are incredibly strong and very good at tree climbing, so they can carry prey double their own weight 10 metres up a tree where bigger cats can't reach them.

Lions use stealth and camouflage too, but unlike other cats they are team players. They live and hunt in family groups called prides. Some pride members do the chasing, while others lie in wait to ambush. Prey is tripped and grabbed by paws and claws, then it's killed by suffocation: one of the team crushes the animal's windpipe, or closes off its mouth and nose with a bite like a giant staple gun. Working together, lions kill zebra and wildebeest and can even tackle hippos and elephants.

LUNCH

DINNER

TEAM LION

CHASE THE ZEBRA

AMBUSH THE ZEBRA

THEN TRIP AND ...

LETHAL SPEED

Cheetahs have the same stabbing and flesh-slicing teeth as the other big cats, but they don't have their size and strength. What they have instead is speed. A cheetah has a light, skinny body, long legs and a spine that acts like a spring, so it can bound along covering more than 8.5 metres with a single stride. Instead of claws kept razor-sharp by sheathing, cheetah claws are always out and used like the spikes on running shoes. All this makes cheetahs the fastest four-footed hunters on land, and twice as fast as the fastest human sprinter. They can reach 96 kph in a few seconds and bring down gazelles – tripping them up, then killing them with a bite to the neck. But cheetahs can't keep running. If they don't catch up with their prey in about 60 seconds, they overheat and have to give up.

TOP DOGS!

Big cats may have big teeth and big claws, but they don't actually catch food very often. Just one in three lion hunts results in a kill and cheetahs only kill once in every four hunts. If you want a really efficient killer, you need dogs. Like cats, dogs have highly tuned senses to track down prey and top-class killing tools in their jaws. But unlike cats, dogs' jaws are long, giving them a more powerful bite, and unlike cats, they can run, and keep running.

African wild dogs are about the size of a collie dog. They hunt in packs and catch something almost every time they go out hunting. They can bite harder, for their size, than any big cat, and even outdo their doggy cousin, the wolf, in bite power. But the real secret of wild-dog hunting success is the ability to keep running for hours on end until whatever they are chasing gets tired. Then the pack closes in, with one or two dogs biting the animal's nose to hold it still, and the others simply biting its tummy so that its guts drop out. The pack may start to eat even before their victim is actually dead.

It is the doggedness of wild dogs that keeps them ahead of the cats when hunting. But it can also be their downfall. African wild dogs use up so much energy hunting that if their kill rate drops even a little, they can starve. Cats, however, can stay chilled about a bit of failure.

11

BIG BAD WOLF

The biggest and baddest of all the dog family is the wolf. Wolves are twice the size of African wild dogs and can weigh as much as a fully grown human. They have a powerful bite – second only to African wild dogs, and like their relatives, they are great runners, with a top speed of 64 kph and the ability to keep up a steady trot for 32 km. They travel more than 40 km between kills and spend as much as half of their time on the move.

Wolves often hunt in packs and work together to catch prey. But finding it in the first place relies on their ability to spot tiny clues of scent, sight and sound that show them where food is in any habitat and in any season. Also, unlike cats, wolves can change how they digest their food so they can eat almost anything that comes their way.

Once, wolves were the most successful and widespread predators on land, and were found all over Europe, North America and Asia, from hot deserts to snowy Arctic wastelands. They were so successful that they became very unpopular with humans because they ate sheep and cattle. Wolves now only survive where there are few people to kill them.

In fact, only one predator has a worse reputation with humans than the wolf – the shark.

JAWS

Big predatory sharks are so scary, it's as if they have swum straight out of a horror movie into the real world. Their eyes are super-sensitive to movement, so they can see things that would be a blur to us. They can smell blood in the water from kilometres away and their skin can feel the tiniest ripples made as their prey swims. In addition, little gel-filled pits (the wonderfully named "ampullae of Lorenzini") pick up the electricity in nerves of prey, so they can bite their victim, even if they can't see it.

Sharks are never, ever without their bite. A shark's mouth contains rows of teeth, one behind the other, so that if one tooth falls out, it is immediately replaced. Each tooth is shaped like a triangle for strength and serrated on the edge like a steak knife. When the shark bites, the whole jaw with its hundreds of teeth moves forward slightly out of the shark's mouth, and closes with enough power to cut through flesh and bone like a hot knife through butter.

And, of course, sharks can be very big indeed. The biggest predatory sharks, great whites and tiger sharks, can reach lengths of over 6 metres. These massive sharks can attack and eat almost anything from huge sea turtles to dolphins, but there's one creature in the sea with an even bigger bite...

WOLVES IN THE WATER

Killer whales, or orcas, can be two metres longer than the biggest great white. They have 40 or 50 cone-shaped and backward-pointing teeth, each one 7.6 cm long, and a mouth big enough to bite a seal in two.

But it isn't just their big mouths and sharp teeth that make orcas deadly killers. Like wolves, killer whales hunt in packs and they communicate using whistles and clicks to work as a team. This, together with their ability to find prey in deep water using echolocation, gives them lots of different hunting techniques: they can herd fish like underwater sheepdogs; they can tip ice floes to slide resting seals into their jaws; they can surf up beaches to catch baby sea lions in the shallows. Orcas may even have special techniques for killing sharks, flipping them upside down to paralyse them or holding them still so water can't flow over their gills, which stops them breathing.

These may seem particularly gruesome ways to kill your dinner, but crocodiles do something even nastier.

LET'S TWIST AGAIN

A big croc can be over 5 metres long (the biggest on record was just over 7 metres). But in spite of its size, a crocodile can make itself almost invisible by lying still at the water's surface, with just eyes and nose poking out. Then it lunges out of the water to grab its prey with enough power to lift a car. Up to 68 cone-shaped teeth hold its victim tight while it is dragged under the water, where the crocodile can hold its breath for two hours, easily long enough for its prey, such as zebra or wildebeest, to drown. The crocodile then performs the so-called "death roll", clamping its jaws round a body part – like a head or leg – and spinning in the water to twist it off.

THE FASTEST CLAW

Size isn't the whole story when it comes to killing. The animals that pack the most powerful punch could hide under a lion's paw.

Mantis shrimps are about the size of pencil cases. They live in burrows on tropical reefs and their gorgeous colours might fool you into thinking they are decorative and harmless. Far from it. Their two front legs are modified to form either a barbed dagger or a brutal club. The shrimps can shoot these weapons towards their target at the same speed as a .22 calibre bullet, so that even if the mantis shrimp misses, the shock wave it creates will stun its prey.

But the hardest, fastest bite in the animal kingdom belongs to the trap-jaw ant of the forests of the Caribbean and Central America. It can snap its pincer-like jaws shut 2,300 times faster than you can blink your eye. This is so fast that the jaws create a force 300 times the ant's body weight – bad news if you are a caterpillar caught between them. Both ant and mantis shrimp rely on a latch system to give them their power. Muscles pull back the legs or jaws and hold them cocked and ready, like an arrow held in the pulled string of a bow. When the latch releases, all the pent-up energy is released in a fraction of a second, shooting the legs or jaws into motion, like an arrow flying from a bow.

THE FASTEST CLAWS UNDER THE SEA!

TOO FAST FOR ME!

PYOW!!

A VERY NASTY SHOCK

We think of electricity as being a human invention, but animals have been using electricity to carry messages around their bodies and trigger muscles into moving for millions of years. The electricity in your body would run a light bulb, but there's one creature that has enough electrical power to light up a whole house, and it's wired up as a weapon.

The electric eel skulks around at the bottom of muddy rivers in South America, looking rather like an overstuffed grey sock. It has converted its messages-to-muscles system so that three-quarters of its two-metre-long body is devoted to making electricity. Most of the time, the eel generates a gentle electric pulse of just a few volts to create an electric field around itself. Any objects in that field bend the flow of electricity; the eel feels the kinks and uses them to navigate through gloom and mud.

But when it finds a creature that it wants to eat or a larger animal is threatening it, the electric eel can generate 500 volts of electricity — twice as much as the socket in your wall and easily enough to kill a human at close range. This means that the eel can stun small creatures to eat and keep bigger ones, like jaguars and humans, at a safe distance.

← AN OVERSTUFFED GREY SOCK.

WINGED AND DEADLY

So far we've talked a lot about killing on land and in the water, but what about airborne deadliness? Birds of prey are some of the fiercest killers on the planet. They use their beaks for tearing flesh off bones, and their claws, or talons, for killing.

Owls are the ultimate stealth hunters. Their huge eyes can see when there's even the tiniest bit of light. Their ears, set at different heights on either side of their heads, pinpoint the position of any sound. Then the soft feathers of their wings make their flight completely silent, so prey has no idea that it's about to be stabbed to death with needle-sharp talons.

Hawks and falcons are daytime hunters with eyesight up to nine times sharper than ours; a golden eagle can spot a hare on a hillside more than a kilometre away. Airborne death-by-talon stabbing is then delivered in a variety of different ways: eagles ambush from above; hawks chase and snatch, or, most deadly of all, there's the peregrine's "death by dive-bombing". Peregrine falcons dive on their flying prey from a height of several hundred metres, like plummeting missiles. They can reach speeds of 200 kph or more, and experience g forces greater than fighter pilots. They hit their target with a force 25 times more than their own body weight, killing a pigeon instantly by snapping its neck.

LEGLESS KILLING

How can you be deadly if your body is just a skinny tube? This is the problem that snakes have to face and as all 2,700 different species of them are predators, it's clear that evolution has given them some solutions.

One solution is to use your body for strangling. Constricting snakes like pythons and boas simply wrap themselves around their prey and squeeze the life out of it. The other solution is poisoning — and fast. Every species of venomous snake has a different venom, each one a murderous cocktail of 20 or more different chemicals that cause death in many ways: some paralyse victims; some suffocate them or cause heart failure; some just dissolve flesh. All are delivered in a super-fast bite, called a strike, that takes just a fraction of a second. Venom rushes from sacs around the snake's jaw, down grooves in the teeth and into the bite wound. The most advanced snakes, such as vipers, have two hollow, pointed teeth that they use like hypodermic syringes.

Then the only real problem is swallowing. Because snakes don't have teeth for chewing and often kill animals bigger than themselves, their lower jaws part at the front and both jaws are only loosely joined to the skull so that their mouths can open really wide and be stretched slowly over their prey.

TEENY AND TOXIC

Poison is such a great way of killing that many small animals use it too. Almost all of the world's 40,000 species of spider have poison glands linked to a pair of fangs just in front of their mouths. But their success as killers relies on both ends of their bodies — silk made by glands in their bottoms helps to make all kinds of cunning traps, from silken lassos and tripwires, to webs. Spitting spiders combine the skills of both their ends by bending their bottoms forward towards their heads when they are ready to kill. They squirt a sticky silk from their bottoms and venom from their fangs to glue their prey to the spot and poison it at the same time.

The most truly toxic small animals aren't on land but in the sea: box jellyfish. They aren't really jellyfish but members of an ancient group of living things that have been around for 540 million years. They have 24 eyes and don't drift about like jellyfish, but swim fast in very straight lines. There are lots of types of box jellyfish, some as big as a hat, some as tiny as a fingernail and all with tentacles covered in stinging cells, loaded with a mysterious toxin. No one knows exactly what the toxin is made of but it kills big fish in seconds, making them an easy meal for a squishy-bodied box jelly. Scientists studying them think they could be the most toxic deadly creatures on the planet.

ATTACK TO DEFEND

Poison, like any sort of weapon, can be used for defence as well as attack. Rattlesnakes kill small birds and mammals in seconds with their bite, but if something bigger threatens them, they'll rattle the dried-up scales on their tails as a warning; if the animal doesn't take the hint, then rattlesnakes strike to defend themselves. Cobras rear up and spit their poison out through their fangs at the eyes of any large animal threatening to trample them; their aim is amazing and they can even predict how their enemy will move, to blind it with poison every time.

Scorpions are related to spiders but keep their poison at the other end of their bodies — the "sting in the tail". The sharp, curved sting stabs and delivers a dose of poison from the two little sacs joined to it, but usually only in defence.

You don't have to be a predator to use a poisonous sting. Bees are peaceable nectar collectors until something attacks their hive and then they turn into kamikaze pilots, stinging the intruder even though each sting costs the bee her life. Bee venom contains a mixture of chemicals that cause maximum pain. So just a few stings are usually enough to send a hungry honey thief running for cover.

21

EXPLOSIVE DEFENDING

Ants, like bees, live in colonies – each one of which is a big family. They will die to defend their homes and relatives, and carpenter ants have a particularly nasty way of doing it. The bodies of workers contain huge glands filled with gluey poison. When the colony is threatened, workers explode, covering their enemies in goo. Small predators such as other insects are killed by suffocation and bigger ones, like birds, get so gummed up they stop the attack.

When you don't have lots of relatives to protect, suicide is pretty pointless, so bombardier beetles have a more controlled way of using explosives. They keep two chemicals in separate compartments in their bottoms, but when they are threatened they squirt both chemicals out together. The mix is explosive, heating up to boiling point in a fraction of a second and bursting out of the beetle's bottom in a high-speed, scalding, toxic spray. The beetle's aim is good too, as it has little flaps around the spray's exit that direct the mixture right where it's needed. Of course bombardier beetles are pretty small, so a tiny stinging spray isn't going to keep a predator at bay for long, but it's just long enough for the beetle to unfold its wings from their covers and make a nifty airborne getaway.

22

DO NOTHING DEADLY

You don't have to bite, sting or explode to be deadly. You can just sit there doing nothing.

Poison-arrow frogs from South America are quiet little creatures that look like brightly coloured jewels, but are some of the most poisonous animals on the planet. The frogs' colours advertise the fact that their skin contains poison so deadly that you could die from just holding one in your hand. Native people use it to tip their arrows and predators know that the frog's bright colours mean "if you eat me you'll die", and keep well away, without the frog having to do a thing.

The fugu or puffer fish employs the same tactic. Its flesh is packed with poison, making it a very unhealthy meal indeed for marine predators (although Japanese people regard fugu as a delicacy and have been eating it for thousands of years – sometimes dying by eating the wrong bits).

Scientists have found that poison-arrow frogs, fugu and quite a few other toxic creatures contain the same kind of poison – TTX – but animals don't make it themselves. They get it from their natural food, or from bacteria that live in their bodies. Farmed fugu, kept free of microbes, have no TTX and captive reared poison-arrow frogs would be useless for tipping poison arrows.

WRONG PLACE, WRONG TIME

With so many weapons around in the animal world, it's almost inevitable that once in a while they get turned on us humans. Sometimes this happens because we simply get into the wrong place at the wrong time.

Spiders are found in every habitat on earth, from remote rainforests to the corner of the bathroom — it's a good thing that most of them have fangs too tiny to pierce human skin. But there are a few species of spider that can kill a human with their bite and like to hang out in the dark corners of sheds, woodpiles and outdoor toilets (or dunnies as the Aussies call theirs). The Australian redback and its American relative, the black widow, are smaller than a sugar cube but have venom 15 times more toxic than a rattlesnake's and fangs that pierce human flesh. Sydney funnel-web spiders are much bigger — up to 5 cm long — and have a poison that's 200 times more toxic than most rattlesnakes'. Funnel-webs often fall into swimming pools — but as they can easily survive 24 hours on the bottom of the pool, it's just not safe to rescue them!

The spider that has killed the most people is much less famous — it's the Brazilian wandering spider. It's as big as the palm of a hand, has the biggest poison glands of any spider and likes to hide in shoes. What could possibly go wrong?

SNAKE MISTAKE

THE WRONG PLACE AT THE WRONG TIME.

People who work in the fields and forests of poor tropical countries are often in the wrong place at the wrong time when it comes to snakes. They disturb a snake accidentally and, as they are probably not wearing shoes, they get bitten. 100,000 people die every year because of these accidental encounters.

In Tamil Nadu in Southern India, where lots of people get bitten by snakes, the Irula Snake Catchers Industrial Cooperative Society is trying to help. The cooperative members are trained from childhood to capture deadly snakes without getting bitten. Each captured snake is cared for, fed on plenty of rats and frogs and regularly "milked". Milking a snake means getting it to bite a hollow container so that the venom can be collected and used to make anti-venom – medicine that fights the effects of snakebites. Money from the sale of anti-venom pays for shoes, homes, education, health care and snakebite treatment for cooperative members.

Afterwards, all snakes are returned unharmed to the place where they were caught. This isn't just because snakes are the source of valuable anti-venom; they have another important role, killing rats that would eat stores of human food or spoil them with their droppings.

PLAIN STUPID

Humans are not always just unlucky victims. We deliberately hunt and kill animals, and trespass on their homes and territories; then, quite rightly, they fight back. But sometimes we get hurt, not because we meant to threaten an animal, but because we were just plain stupid.

Brown bears – or grizzlies as they are called in North America – can be twice as big as a tiger. Their jaws are extremely powerful and they have paws the size of dinner plates, armed with claws that can be 10 cm long! A brown bear can kill a fully grown moose weighing 800 kg. Black bears are smaller, somewhere between leopard-size and lion-size, but are still easily big enough to bring down a deer. So when you are in bear country you need to use your brain.

People are killed by grizzlies and black bears every year in the US and Canada, because they didn't stick to the rules: maybe they had a ham sandwich in their backpack, or maybe they thought

Never surprise a bear. Make a noise so it can hear you coming and get out of the way.

Never, ever try to get close to a bear — especially not cubs.

Don't keep tasty-smelling food in your tent, or in your backpack, because if a bear wants it, it will get it — you are just a wrapper, and wrappers get ripped.

they could get close to a cute cub and its mother wouldn't mind. Once a person is dead, bears don't like to waste the chance of a good meal, so they may take a few bites of the body.

Using your brain helps to keep you safe from shark attacks too. Sharks mostly don't want to eat humans. We're a bit bony and not as fat as some of their favourite foods, such as seal. But if we are stupid enough to go swimming in dark or murky water where their real prey usually swims, there can be a nasty case of mistaken identity. Sure, the shark will probably spit you out, but even being tasted by a 5-metre great white isn't very good for your health. Learning to speak shark helps — a shark hunching its back at you with its fins down is telling you to "GO AWAY", and if you don't get the message, who can blame the shark for attacking?

YOU *ARE* DINNER

Animals that turn their weapons on us in self defence or by mistake – theirs or ours – are nothing like as scary as those that attack us just because we make an easy meal.

Tigers, leopards and lions have all been known to eat people and some of them have become famous for it. The tigress known as the Champawat man-eater roamed the border of Nepal and India from 1903 to 1911 and polished off 436 human meals-on-legs; the man-eater of Panar, a leopard, ate 400 people before it was shot in 1910, and the Njombe lion pride from Southern Tanzania ate over a thousand people between them from 1932 to 1947.

Wolves, although they are not as fierce as their reputation, also have an occasional snack of human, although being much smaller than lions or tigers, their favourite human dinner is a 3 to 11-year-old child left playing outside alone, or sleeping in the open air. In 1878 in just one state in India, 624 people – mostly children – were killed and eaten by wolves.

Many of the famous man-eaters of the past put humans on their menu because they were injured. Humans, being slow-moving and plentiful, were the only thing they could still manage to catch. But today, healthy big cats and wolves are eating people because there are fewer natural prey and we humans are everywhere!

The Niassa Reserve in Mozambique is one of the largest in all of Africa where 30,000 people live side by side with lions. In the last thirty years, there have been 75 attacks by lions. Eleven people have been killed since 2000.

Scientists and local people are working together on the Niassa carnivore project to try and sort this out. They found that attacks happen when lions follow their favourite prey — bush pigs and warthogs — into fields of crops. There, the lions come across people sleeping outside, or walking alone in the dark. Expecting a lion not to attack an unprotected human is a bit like leaving the sunday roast on the kitchen floor and expecting your dog not to lick it.

Building fences and stronger huts keeps lions, and their prey, out of crops and homes; and leaving some natural habitat undisturbed so lions' prey can thrive cuts the risk of attacks drastically.

DEADLY ... AND DEAD?

Animals that are armed and dangerous are not very popular. Humans just get rid of any animal that is seen as a threat. In Britain, wolves were wiped out in the eighteenth century, and in Africa, lion numbers are falling fast because people who have been attacked by lions kill lions in return.

But killing deadly animals is a poor solution because they can be very useful. Big predators like lions, tigers, wolves and leopards may be dangerous, but without them, the number of animals that eat crops would soar. Snakes may bite us, but they do a fantastic job of killing rats that would otherwise be munching their way through our food.

Predators keep the natural balance in the sea as well. All over the world, shark numbers are dropping, as they are caught in nets or have their fins chopped off to make into soup. But without sharks, the balance of many marine habitats is upset and fishing nets come up empty.

Predators have another value too. They attract tourists – and tourists bring jobs and money. Who would think of going on an African safari holiday without lions? It's clear that deadly animals need an image make-over, for our sakes and their own.

LIFE-SAVING POISONS

Animals with deadly poisons can be useful. Their venoms are a mixture of chemicals, each one of which may have an effect on a human body and could be the source of new medicines.

Scientists have taken a chemical out of scorpion toxin that they've found sticks to cancer cells like glue. By attaching a coloured dye to the chemical from the scorpion venom, and injecting it into cancer sufferers, doctors can see where every little bit of a tumour is – and remove it. Extracts from spider venom are being used as a treatment for stroke victims. In small doses cobra venom helps to relieve the symptoms of arthritis and scientists are also investigating bee venom, as it too may have arthritis-busting action.

One of the most powerful venoms is found in tropical cone shells. Like box-jellyfish toxin, the venom is super-poisonous and stops fish dead in their tracks. Cone-shell venom is unique: it's made of hundreds of different chemicals, some of which have a very precise effect on particular kinds of human nerve and brain cells. What's more, these chemicals – unlike those from spider or snake venom – are really easy to make. Cone-shell toxin is now helping scientists to make new kinds of painkillers, and to investigate how our brains work.

LIVING WITH DEADLY

In our modern world, where there are simply so many people, it's important to remember that animals, even the armed and deadly ones, have just as much right to their place on the planet as we do. It's up to us to keep ourselves and those animals safe, and, as we've seen, that has all sorts of benefits.

We should respect the power of deadly animals, and understand that if we don't behave sensibly they can hurt us. But that doesn't mean being afraid of them for no reason. Every time you go into the sea, you are 132 times more likely to drown than you are to be killed by a shark.

And we share our homes with a very dangerous descendant of the wolf — the domestic dog, which kills about 25 people in the US every year.

When we enter a wild animal's world, we shouldn't expect it to be either a monster or a best friend. It is just itself and if it attacks us with teeth or claws, stings or poisons, it's just doing what millions of years of evolution have kitted it out to do. It doesn't have a choice. Humans are the ones who can make choices, and we can choose to share our world, quite safely, with even the very deadliest of creatures.

INDEX

GLOSSARY

Birds of prey birds such as hawks, falcons, buzzards, eagles and owls who hunt for prey to kill and eat.

Echolocation using echoes of your voice to find your way around. Bats, dolphins and some whales do it. Some humans with visual impairment can do it a bit too.

Evolution the way that living things change over many generations, so that new kinds of living things come into being.

g force the force on an object when it accelerates. The greater the acceleration, the greater the g force. (Swing a ball on a string round your head and you'll feel the g force pulling.)

Pack the name for a group of wolves, wild or domestic dogs hunting together.

Predator an animal that hunts and kills other animals for food.

Prey an animal that is hunted and killed by another animal for food. Some animals can be both predators and prey. Seals are predators of fish, but are the prey of polar bears.

Pride the name for a family group of lions, who defend a territory, live and hunt together.

BITING!

FEED SUFFOCATING!

CRUSHING! UGH!

POISONING! UGH!

ABOUT THE AUTHOR

Nicola Davies is an award-winning author, whose many books for children include *The Promise*, *A First Book of Nature*, *Big Blue Whale*, *Dolphin Baby*, and *The Lion Who Stole My Arm*. She graduated in zoology, studied whales and bats and then worked for the BBC Natural History Unit. Visit Nicola at **www.nicola-davies.com**

About *Deadly* Nicola says, "Living with animals that are cute and harmless is easy, but we need to share our world with the creatures that can bite and sting too."

ABOUT THE ILLUSTRATOR

Neal Layton is an award-winning artist who has illustrated more than sixty books for children, including the other titles in the Animals Science series. He also writes and illustrates his own books, such as *The Story of Everything* and The Mammoth Academy series. Visit Neal at **www.neallayton.co.uk**

Of illustrating *Deadly* he says, "It was really good fun drawing all these dangerous animals, and now I'll know what to do if I ever meet a shark who puts its fin down and hunches its back at me…!"

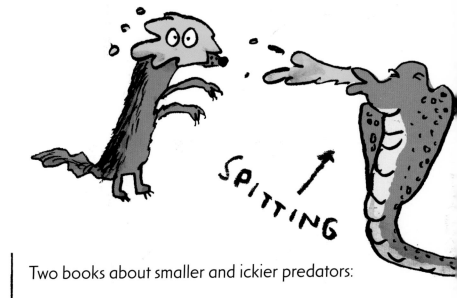

SOURCES

This book is about deadly animals by a writer who has met them all:

Predators: The World's Most Lethal Animals by Steve Backshall (Orion Childrens, 2013)

Two books about smaller and ickier predators:

Explore the Deadly World of Bugs, Snakes, Spiders, Crocodiles by Barbara Taylor, Jen Green, John Farndon and Mark O'Shea (Armadillo Books, 2013)

Black Widows: Deadly Biters by Sandra Markle (Lerner, 2011)

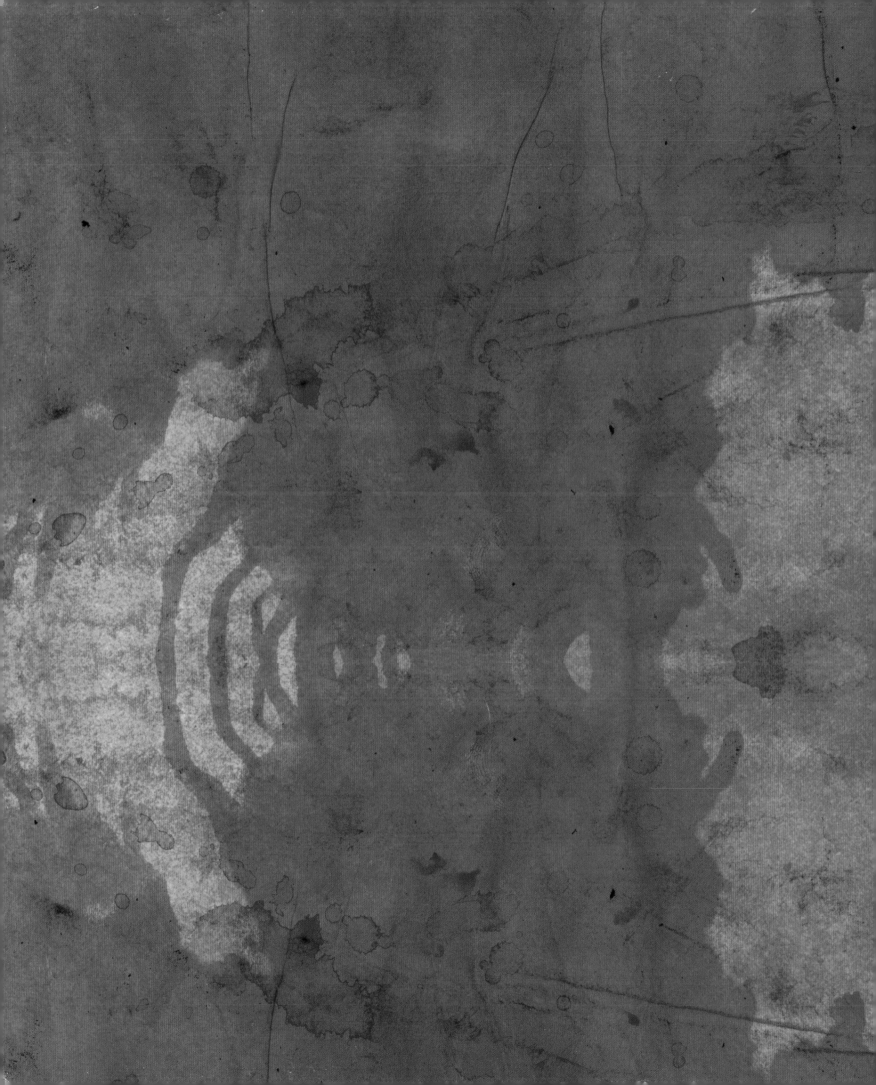

ANIMAL Science

HOW AND WHY ANIMALS DO THE THINGS THEY DO.

ISBN 978-1-4063-5663-2

ISBN 978-1-4063-5665-6

ISBN 978-1-4063-5742-4

ISBN 978-1-4063-5747-9

ISBN 978-1-4063-5748-6

ISBN 978-1-4063-5664-9

If you enjoyed this book, why not collect them all!

Available from all good booksellers

www.walker.co.uk